Weather and Climate

Reader

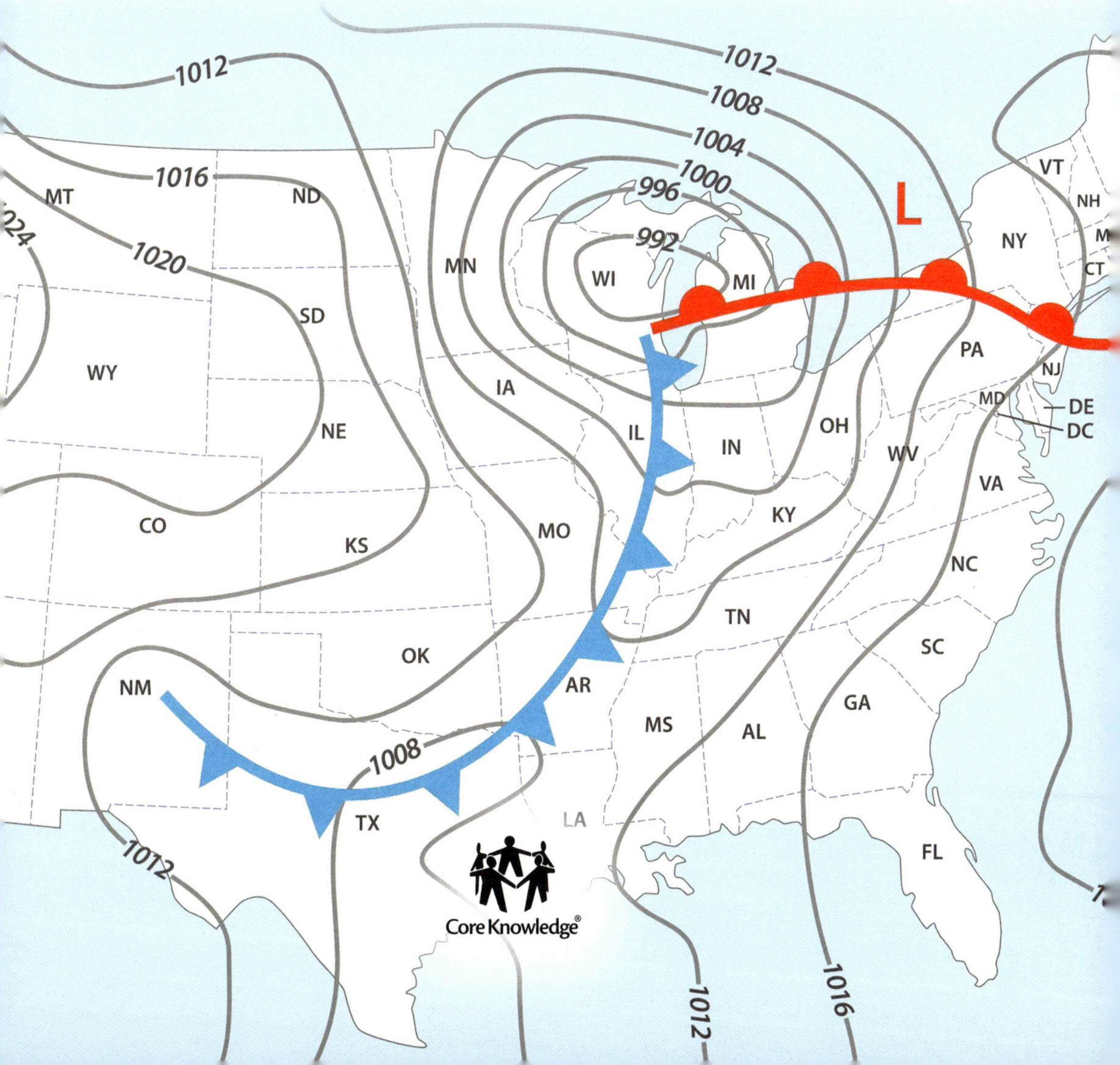

Core Knowledge®

www.coreknowledge.org

Printed in Canada

ISBN: 978-1-68380-514-4

Weather and Climate

Table of Contents

The Atmosphere and Air Pressure

Chapter 1

Suppose a friend asked you, "What is the weather like today?" What would you say? How would you know? You might say, "It's cold and raining."

Big Question

What is the atmosphere, and what is weather?

Weather is what the air outside is like at any one time and place. The weather can be clear and warm, cold and rainy, or dry and windy.

Vocabulary

weather, n. what the air outside is like at any given time and place

atmosphere, n. the layer of air that surrounds Earth

You can observe the weather. You can look outside. You can collect rain in a container and measure it. You can find out which way the wind is blowing.

But what if your friend asked you, "What causes different kinds of weather?" What would you say to that?

To understand what causes weather, we need to learn about the **atmosphere**. The atmosphere is the layer of air that surrounds Earth. The atmosphere is about seventy miles thick. Most weather happens in the part of the atmosphere closest to Earth's surface.

Do you see what looks like a thin, glowing blue line? That and the clouds you see are parts of Earth's atmosphere.

Air Is Matter

The air that makes up our atmosphere is a mixture of gases. Gas is a state of matter. Air contains mostly nitrogen gas, but it also contains oxygen gas. Other gases such as water vapor and carbon dioxide are in the air too. Though you never really see the air, it surrounds you every day and helps you to survive.

Think About Matter

Think about water. When you drink water, it's liquid. If you put a tray of water in the freezer, it turns to a solid, ice. If you leave a tray of water out in the hot sun, it evaporates into a gas called water vapor. Liquid water, ice, and water vapor are all examples of different states of matter of water.

Like all matter, the gases in the air take up space. These gases of the atmosphere do not float out into space, though. That is because the force of gravity pulls air down toward Earth's center.

What do you think causes turbines like these to spin? The pushing force of the wind provides evidence that air is made of matter, which can move and transfer motion energy to objects.

Air Pressure

Gravity pulls on air just like all other matter. That means that air constantly pushes against everything on Earth's surface. The weight of air pressing on objects is called **air pressure**.

> **Vocabulary**
>
> **air pressure, n.** the weight of air as it presses on objects below or within it

Air pressure is less as you travel upward and away from the surface of Earth. An object very high in the air experiences less air pressure than another object at sea level.

That is because there is less air higher up in the atmosphere to press on things. Air pressure changes depending on the amount of matter in it. If more matter, air, is above and surrounding you, then the air pressure will be greater. Air pressure is one factor that affects weather.

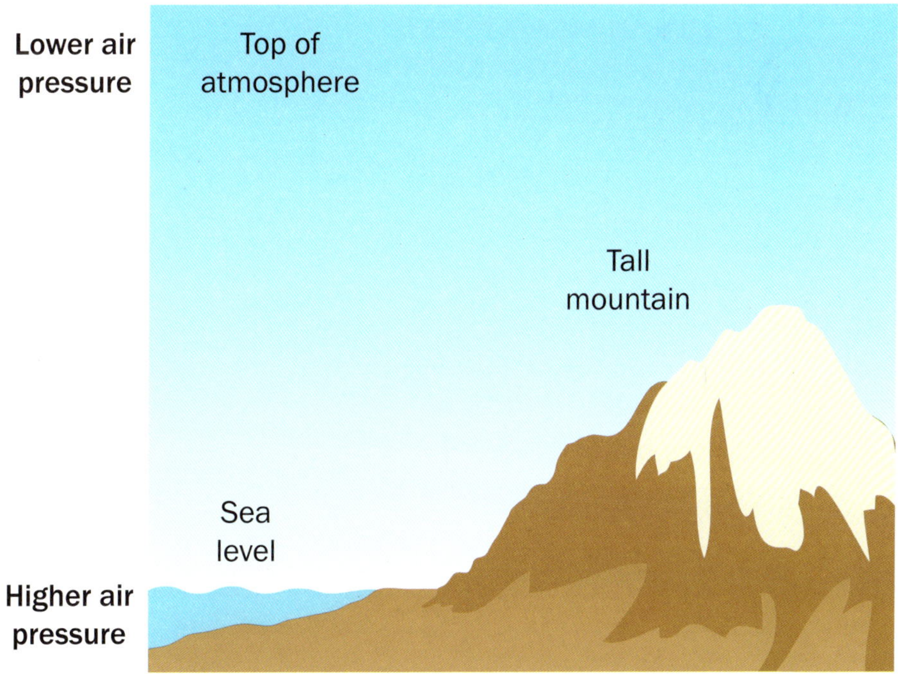

Sea level is the level of the surface of the ocean where water meets the land.

Changes in Air Pressure

Energy from the sun warms Earth's surface. It also warms the atmosphere. Do you know what happens to air when it warms? It spreads out. When air cools, it gets more compact, filling less space. Changes in air temperature and air pressure cause many of the conditions that we know as weather.

When air pressure is high, the skies are usually clear and sunny. Low air pressure in an area is often associated with clouds, rain, or even snow.

Think About Matter and Temperature

Temperature is a measurement of how hot or cold something is. Typically, when matter increases in temperature, it begins to spread out. For example, when liquid water is boiled to become water vapor, the tiny particles spread out so much that it becomes hard for anyone to see them.

How would you describe the weather? Do you think the air pressure in this area is high or low?

Water in the Atmosphere

Chapter 2

What do you think of when you think of nice weather? Do you imagine a dry, warm, sunny day? Or maybe you picture a clear and cool day. You might even think of a day with warm, gentle rain. Sunshine is necessary for living things. But rain is important too.

Big Question

How does water move into and out of air?

There is water on Earth's surface. There is water in the oceans and lakes. There is water in the atmosphere, too, and even underground. All this water moves from Earth's surface to the atmosphere and then back again all the time. This water takes different forms as it moves. The movement of water from Earth's surface to the atmosphere helps all living things survive.

If a place has many green plants, that is a sign that it probably rains often there.

Water Falls to Earth as Precipitation

One of the gases that occurs in our atmosphere is **water vapor**. Water in its gas form is called water vapor. Sometimes air contains a lot of water vapor. At other times it contains less. When water vapor gas cools, it may change to tiny droplets of liquid water. This liquid water may fall from the sky. This is **precipitation**. Precipitation can take the form of rain, snow, sleet, or hail.

Vocabulary

water vapor, n. the gas form of water

precipitation, n. water that falls from the sky in the form of rain, snow, sleet, or hail

Rain is drops of liquid water. Snow is frozen water in the form of ice crystals. Sleet is tiny frozen pellets of water. Hail is icy, round balls.

The type of precipitation that falls depends on air temperature. Rain falls when the temperature is above freezing. When it is below freezing, sleet and snow fall. Hail forms high in the atmosphere where it is very cold. It hits the ground before it can melt. Hail can fall any time of year.

Several inches of snow can fall when there is enough water in the atmosphere and the temperature is low enough.

Water Rises into the Atmosphere When It Evaporates

You know that water falls from the sky. But how does it get into the air? Water moves in a cycle from Earth's surface to the air and back again.

Energy from sunlight causes water on Earth's surface to **evaporate**. That means that it changes from a liquid to a gas. This gas, water vapor, then rises and becomes part of the atmosphere.

> **Vocabulary**
>
> **evaporate, v.** to change from liquid to gas
>
> **humidity, n.** a measure of the amount of water vapor in the air

Sometimes there is a lot of water vapor in the air. At other times air contains less water vapor. Scientists can measure the amount of water in the air. **Humidity** is the measure of how much water vapor the air in a place contains.

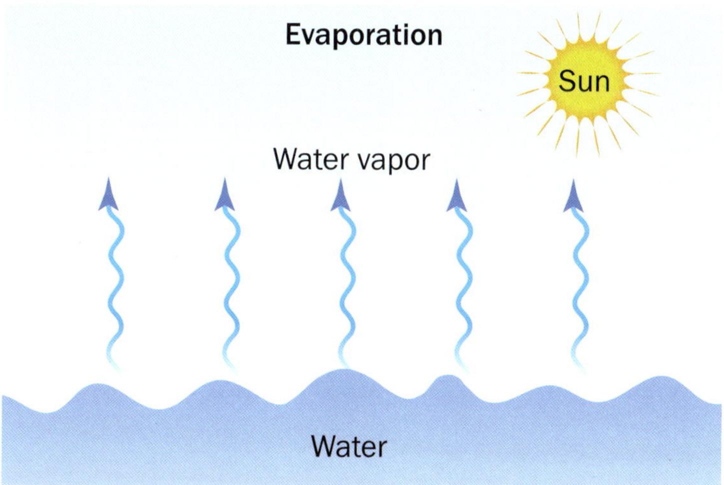

Much of the water vapor in Earth's atmosphere comes from the evaporation of ocean water.

> **Word Parts**
>
> Look at the word *evaporate*. If you break it into parts, what do you notice? It contains the word *vapor*.

Water Vapor Condenses to Form Liquid Water

Water vapor rises into the atmosphere. Air high up in the atmosphere is cooler than air near the ground. This cool air causes water vapor to **condense**, or turn back into liquid. Tiny droplets or ice crystals come together to make the clouds we see when we look at the sky.

Vocabulary

condense, v. to change from gas to liquid

When enough tiny droplets collect in a cloud, they become too heavy to stay in the sky. They fall down to the ground as precipitation. Not all clouds produce precipitation, though. You might see thin, wispy clouds or puffy and white clouds when the weather is fair.

Which kind of cloud will produce precipitation? Which kind looks like it contains more condensed water?

Think About Condensation

Condensation occurs on the ground too! The water you might find on the grass in the morning is known as *dew*. As night falls and light from the sun no longer heats Earth's surface, the surface temperature lowers. At a certain point, known as the *dew point*, water condenses on the surface of objects, such as grass and windows.

Wind

Chapter 3

Look outside. Can you see leaves rustling or tree branches swaying? Or maybe you can see a flag flapping back and forth. When you see this, you know that air is moving. We call it **wind**.

Big Question

What is wind?

Vocabulary

wind, n. the movement of air

Wind is the movement of air. Sometimes the air outside barely moves. Little or no wind blows. At other times, air moves slowly. Wind can blow gently. And sometimes, air moves quickly and forcefully. Wind can blow hard.

You can't see the wind, but you can see evidence of it. You can feel wind when it cools your skin or lifts your hair. Sometimes you can hear wind too.

You can feel evidence of wind when you are outside.

Can wind be considered a force? Is it a push or a pull?

Air Masses

A large body of air in the atmosphere is called an **air mass**. Some air masses are warm. Others are cool. Warm air masses rise in the atmosphere. Cool air masses sink. Air masses form when a large body of air stays in contact with part of Earth's surface. They take on the temperature and moisture of that part of Earth's surface.

Vocabulary

air mass, n. a large body of air in the atmosphere

When a high-pressure air mass comes in contact with a low-pressure air mass, the high-pressure air moves to the low pressure. This movement creates wind. This is like when you blow up a balloon and release it to the outside air. The air inside the balloon has higher pressure. It makes the balloon fly around powered by wind.

Hot-air balloons rise because the air inside them is heated. The air inside the balloon is warmer than the air outside.

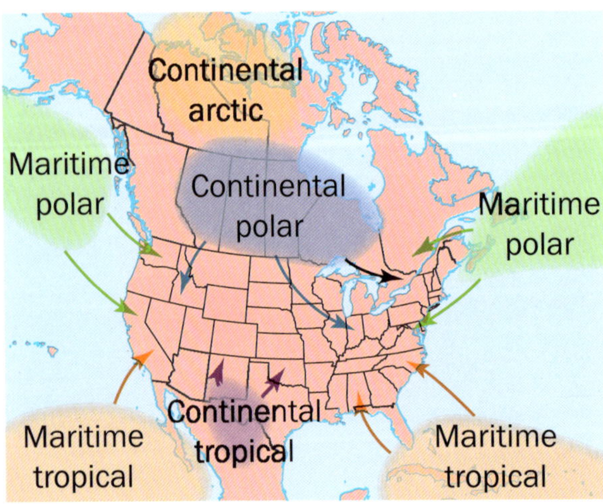

There are different kinds of air masses. The word *maritime* here means over water. The word *continental* means over land. *Polar* refers to the cold air above polar regions. *Tropical* refers to the warm air above tropical regions.

Wind Changes Speed

You know that you can observe wind. But did you know that you can also measure wind? One way to measure wind is to find out how fast it is blowing. **Wind speed** is how fast wind blows over a certain distance and time.

Vocabulary

wind speed, n. a measure of how fast wind blows

Many things affect wind speed. Wind increases when strong high pressure meets strong low pressure or when strong low pressure meets strong high pressure.

Temperature affects wind speed too. There is often faster wind during the day because the sun heats Earth's atmosphere and surface. You can feel this heat if you walk on a sidewalk in bare feet on a sunny day.

Scientists use instruments to measure wind speed. The cups of the instrument move around and around in response to wind speed. That way scientists can tell how fast the wind is moving.

This tool measures wind speed. It is called an anemometer. Wind pushes the cups and causes them to spin.

Wind Changes Direction

You can also tell the direction the wind is blowing. **Wind direction** is the direction from which wind blows. A northerly wind blows from the north. A westerly wind blows from the west. Wind blows from areas of high pressure to areas of low pressure. Wind socks are objects that fill up with air to show which way the wind is blowing. Wind vanes are objects that turn in the direction the wind is blowing.

Vocabulary

wind direction, n. the direction from which air moves when wind blows

prevailing winds, n. regular patterns of winds that blow from one direction

Winds change, but they blow in regular patterns. Regular patterns of wind are called **prevailing winds**. Prevailing winds are winds that blow mainly from one direction.

Understanding prevailing winds helps scientists predict weather patterns. Wind brings changes in the weather. Wind pushes clouds and air masses from one place to another.

Scientists use tools such as wind vanes and wind socks to find wind direction. From which direction is the wind blowing in these pictures?

Using Weather Data to Predict Weather

Chapter 4

A weather forecast tells what the weather will be like for the next few days. It tells what the temperature will be each day. It tells whether precipitation is likely. But where does this information come from?

Meteorologists are scientists who study weather conditions. They collect **data** about weather. They look for patterns in the data. Often, they use the data and computers to find patterns and predict weather in the near future. It is not possible to make a perfect prediction. But collecting data helps meteorologists make predictions that are accurate enough to be useful.

Big Question

What do meteorologists do?

Vocabulary

meteorologist, n. a scientist who studies weather conditions and patterns

data, n. information that is observed or measured and recorded

Word to Know

When you make a *prediction*, you say what is likely to happen.

Computers display data that help meteorologists predict weather.

Meteorologists Collect Weather Data

Meteorologists have tools that help them collect data. The tools are used to take measurements of different types of data.

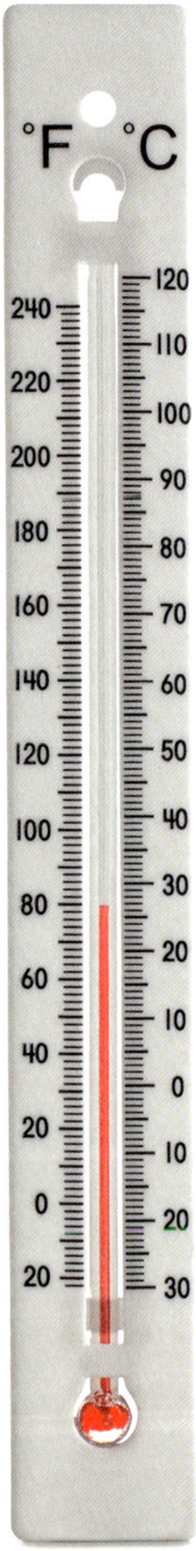

Air temperature is measured with a thermometer.

An anemometer, with cups that spin, measures wind speed. A wind vane shows the direction from which the wind is blowing.

A hygrometer measures humidity, or how much moisture is in the air.

Air pressure is measured with a barometer.

The amount of rainfall is measured with a rain gauge.

A weather station contains several instruments that collect weather data.

Meteorologists also use weather stations to collect information. The Automated Surface Observation System (ASOS) has many stations in the United States. These stations automatically measure temperature, wind speed and direction, precipitation, humidity, and air pressure. They report the weather about every twenty minutes. This helps meteorologists know what the weather is like all over the country at any given time.

Meteorologists Display Weather Data

Meteorologists collect and record data every hour, day, week, and month. Then they organize these data so that they can see patterns and make predictions. Meteorologists use different methods to organize data.

Tables: Weather measurements such as of temperature, humidity, and air pressure are taken many times each day. One way to organize the data is in tables.

	Monday	Tuesday	Wednesday	Thursday	Friday	Saturday	Sunday
12:00 p.m.	63°F	66°F	70°F	57°F	59°F	62°F	64°F
1:00 p.m.	67°F	70°F	74°F	60°F	62°F	65°F	68°F
2:00 p.m.	70°F	73°F	77°F	62°F	65°F	67°F	72°F
3:00 p.m.	73°F	78°F	81°F	65°F	68°F	70°F	75°F
4:00 p.m.	75°F	80°F	83°F	66°F	70°F	72°F	78°F
5:00 p.m.	76°F	81°F	85°F	68°F	71°F	74°F	81°F

What pattern do you see in the daily temperature this week?

Think About Temperature

The table above shows the temperature in degrees Fahrenheit, shown as °F. Fahrenheit is a temperature scale used mostly in the United States. Temperature can also be shown as degrees Celsius, written as °C. Celsius is the temperature scale used in the metric system.

Graphs: A bar graph is a helpful way to see a pattern. It can display how a measurement increases or decreases over time. The graph below shows the high and low temperatures for each day. Which day was warmest? Which day was coolest?

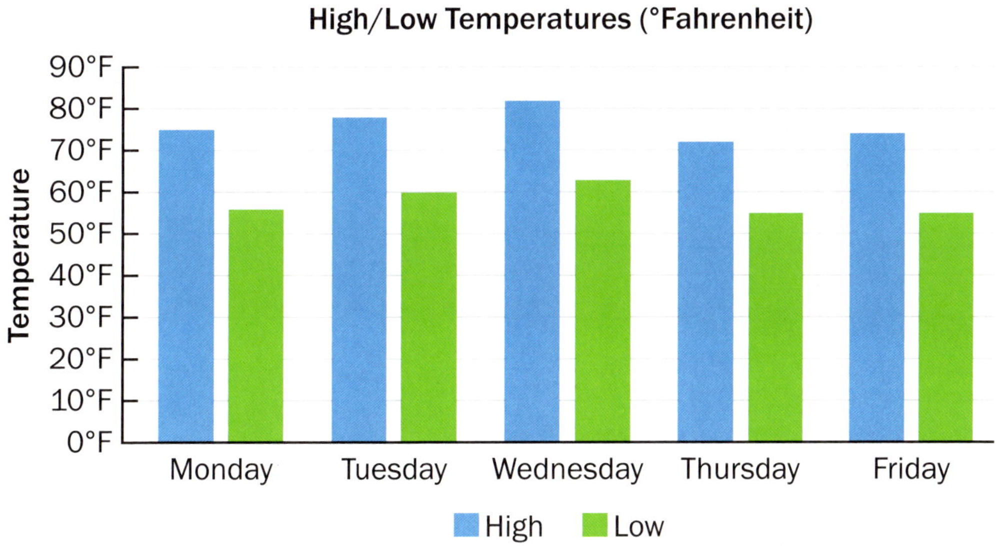

Maps: Weather is what the air outside is like at any given time and place. It is helpful to display weather information on a map. Then people can see weather conditions in other places too.

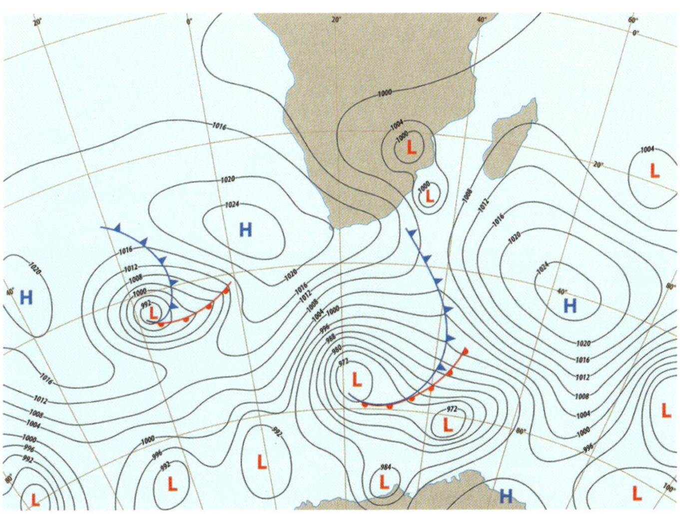

Maps help meteorologists see weather patterns across a large area. This map shows a lot of weather data over an area that is mostly water.

Data Displays Show Patterns and Help with Predictions

Meteorologists draw air masses and wind direction on maps. Air masses move from high pressure to low pressure. The place where two different air masses meet is called a **front**. Cooler weather is behind a cold front line on a weather map. A warm front brings warmer weather. Look back at the table on page 16. What kind of front moved in on Wednesday night?

> **Vocabulary**
>
> **front, n.** the place where two air masses meet

When air masses move from place to place, they carry different kinds of weather with them. Weather is often most active at the fronts. Knowing how and where air masses move allows meteorologists to predict what the weather will be like days in advance.

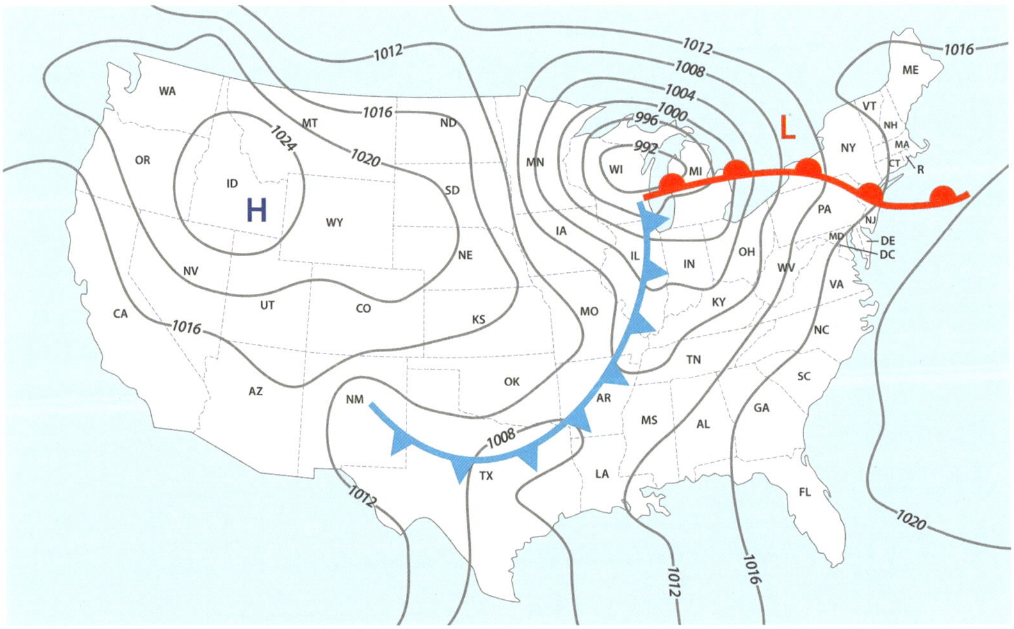

The blue line on a weather map shows a cold front. The red line shows a warm front. The blue triangles and red half-circles indicate which direction the front is moving.

Patterns of Weather: Seasons and Climate

Chapter 5

What season is it now? How has the weather changed since the last season? How will it change between this season and the next one? You have probably noticed patterns in each season. You know that temperatures become warmer or cooler depending on the season. You know that some seasons have more rain or snow.

Big Question

What is the difference between weather and climate?

Average Temperatures (°F) Omaha, Nebraska		
Month	**Low**	**High**
January	13°F	32°F
February	19°F	38°F
March	29°F	51°F
April	40°F	64°F
May	51°F	73°F
June	61°F	82°F
July	66°F	86°F
August	64°F	84°F
September	55°F	76°F
October	43°F	65°F
November	29°F	48°F
December	17°F	35°F

What patterns do you see in this data?

Weather changes from day to day. However, if you look at weather data for one place over a whole year, you can see gradual increases and decreases in temperature and precipitation as the seasons change.

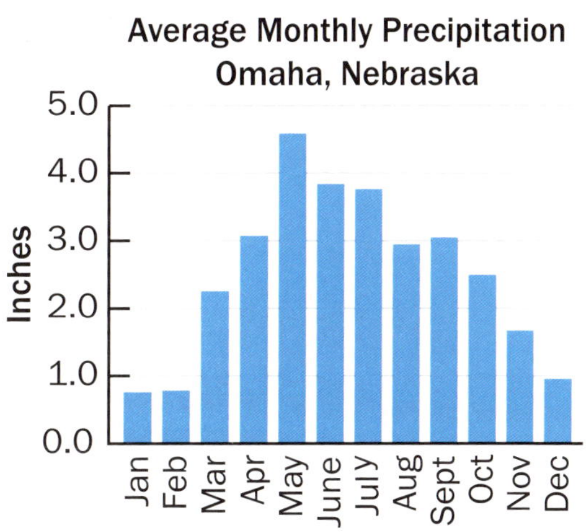

Weather Occurs in Seasonal Patterns

Think About the Causes of Seasons

Earth has an imaginary straight line through the North and South Poles called its axis. Earth is tilted on its axis. As Earth follows a path around the sun each year, the North Pole points away from the sun for part of the year. And the North Pole is tilted toward the sun for part of the year. This means places on Earth are also tilted away from or toward the sun.

Parts of Earth that are tilted toward the sun receive more direct sunlight and have more hours of daylight. Parts that are tilted away from the sun receive less direct sunlight and have fewer hours of daylight. The amount of sunlight affects temperature. It also results in seasonal weather.

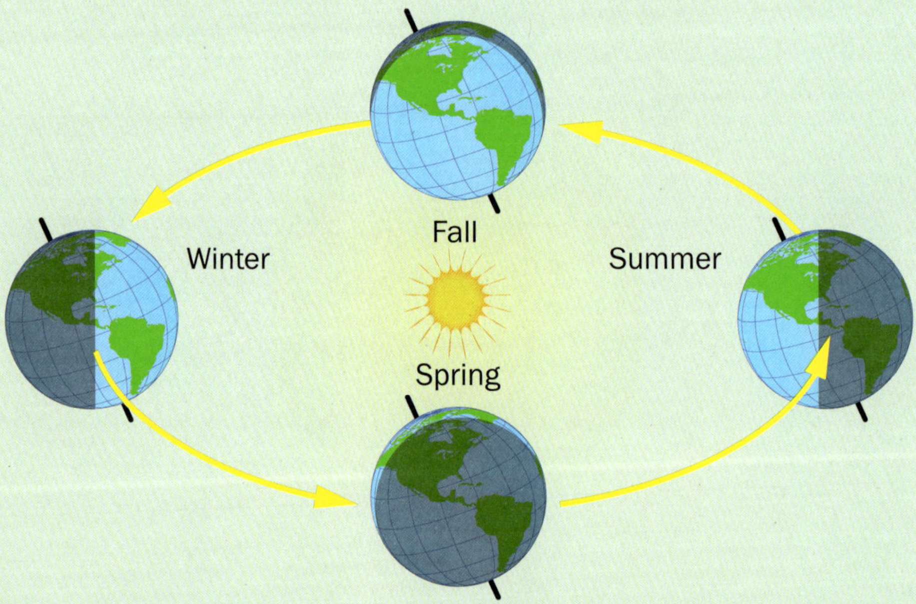

When the North Pole is tilted toward the sun, it is summer in the Northern Hemisphere. But at that same time, the South Pole is tilted away from the sun. It is winter there.

Word to Know

Seasonal means related to the seasons (summer, fall, winter, or spring).

This table shows data for Minneapolis, Minnesota. It shows when the sun rises and sets on the first day of each month. It also shows average high and low temperatures. Can you explain the pattern you see? Are days with more sunlight warmer or cooler? Why?

	Sunrise	Sunset	Average high temperature (°F)	Average low temperature (°F)
January 1	7:51 a.m.	4:42 p.m.	24°F	8°F
February 1	7:32 a.m.	5:21 p.m.	25°F	9°F
March 1	6:50 a.m.	6:00 p.m.	34°F	18°F
April 1	6:53 a.m.	7:40 p.m.	50°F	31°F
May 1	6:02 a.m.	8:18 p.m.	65°F	44°F
June 1	5:29 a.m.	8:52 p.m.	74°F	54°F
July 1	5:30 a.m.	9:03 p.m.	83°F	63°F
August 1	5:58 a.m.	8:39 p.m.	83°F	64°F
September 1	6:35 a.m.	7:49 p.m.	77°F	59°F
October 1	7:11 a.m.	6:53 p.m.	65°F	45°F
November 1	7:51 a.m.	6:01 p.m.	50°F	34°F
December 1	7:31 a.m.	4:33 p.m.	32°F	24°F

Think About Patterns

Read the data in the table one column at a time. Start with the Sunrise and Sunset columns. As you read from one value to the next down the columns, do the times continue to get earlier? Do they continue to get later?

Then look for patterns in the temperature columns. As you read from one value to the next down the columns, do the temperatures continue to get warmer? Do they continue to get cooler?

Climate Is the Pattern of Weather over Many Years

You have learned that weather is what the air outside is like from day to day. **Climate** is the weather patterns in a place over a long period of time. If you compare weather data for a place year after year, you will see that the patterns usually repeat.

Vocabulary

climate, n. the weather patterns in a place over a long period of time

Look at the tables. They show weather data for three different years in two different places. San Jose, Costa Rica, has a warmer climate. Fairbanks, Alaska, has a colder climate.

San Jose, Costa Rica						
	2014		2015		2016	
	Average high temperature	Average precipitation	Average high temperature	Average precipitation	Average high temperature	Average precipitation
Winter	73°F	less than ¼ in.	73°F	less than ¼ in.	75°F	½ in.
Spring	76°F	3 in.	75°F	1 in.	77°F	3 in.
Summer	75°F	4 in.	75°F	5 in.	75°F	6 in.
Fall	74°F	9 in.	75°F	8 in.	73°F	6 in.

The patterns of weather in San Jose, Costa Rica, make up the climate of the area.

Can you find patterns in each table? Try to answer these questions:

- What is the wettest season in San Jose?
- What is the driest season in San Jose?
- What is the coldest season in Fairbanks?
- What is the warmest season in Fairbanks?
- What is the best time of year to visit San Jose if you like warm, dry weather?

Fairbanks, Alaska						
	2014		2015		2016	
	Average high temperature	Average precipitation	Average high temperature	Average precipitation	Average high temperature	Average precipitation
Winter	1°F	6 in.	0°F	7 in.	4°F	6 in.
Spring	32°F	5 in.	37°F	4 in.	38°F	5 in.
Summer	58°F	5 in.	60°F	5 in.	61°F	4 in.
Fall	26°F	7 in.	27°F	8 in.	25°F	6 in.

The patterns of weather in Fairbanks, Alaska, make up the climate of the area.

Earth's Climates Change over Time

At times in Earth's history, climates in some places were different from what they are today. For example, ice ages were long periods of time when temperatures remained very cold in certain places. Large areas of Earth's surface were covered with snow and glaciers. Temperatures on Earth warmed between ice ages.

Look at the graph. It shows how Earth's climate has changed from 800,000 years ago to today. The 0 at the end of the graph is today. The taller the line, the warmer the climate. The shorter the line, the colder the climate.

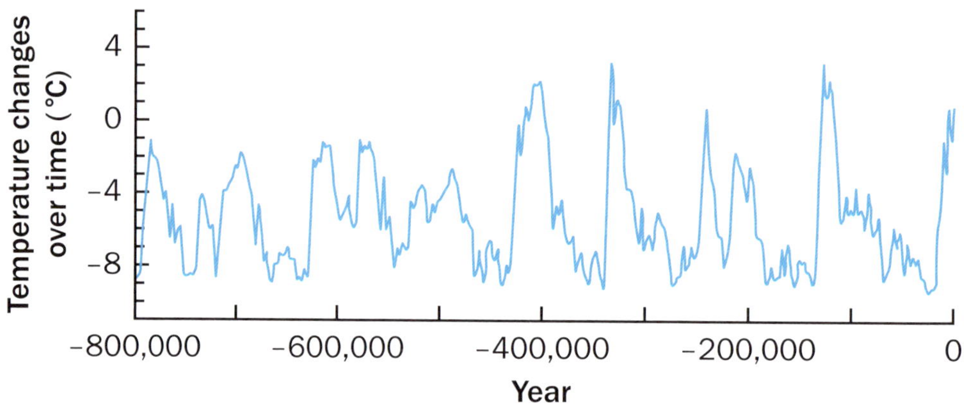

Relatives of elephants known as mammoths lived in North America, Europe, and Asia. They were covered in long hair to protect them from the cold. Mammoths were adapted to a cold climate.

Extreme Weather

Chapter 6

On many days of the year, the weather is calm. Sometimes there is rain or snow. Sometimes it gets very hot or very cold. But the weather usually doesn't keep people from going to work or school. It doesn't usually cause changes or disruptions in a community.

Big Question

What are extreme weather hazards?

Vocabulary

hazard, n. a dangerous condition that can cause damage

However, there are times when weather can be extreme. These weather events usually happen suddenly. They can lead to **hazards**, or dangerous conditions, that threaten safety. Sometimes people have time to prepare in advance for extreme weather. But other times, they do not. It is important to plan for these events, even though they do not occur often.

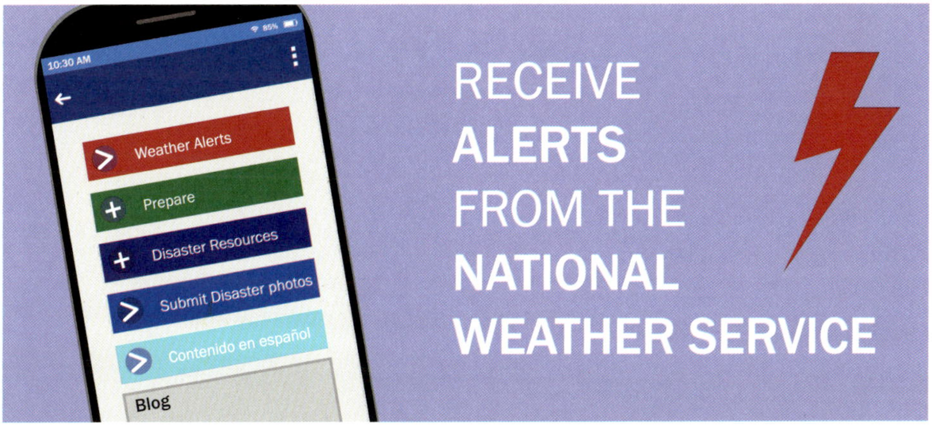

Many phones today can receive weather alerts from the federal, state, or local government. These alerts warn people when bad weather may be coming.

Thunderstorms Cause Lightning and Flooding

You may have experienced a thunderstorm. A thunderstorm is a rainstorm that is accompanied by thunder, lightning, wind, and heavy rain. It is usually over quickly. But some thunderstorms are very strong and can cause serious damage.

Strong thunderstorms may have frequent lightning. Lightning is electricity that travels between clouds, from clouds to the ground, or from the ground to the clouds! It can strike objects, such as trees and houses. When it strikes these objects, the lightning can cause fires. Lightning can also strike people and cause serious injuries.

Severe thunderstorms can also produce large amounts of rain that lead to flash flooding. Flash flooding is a flood that happens with little warning. These types of floods can be powerful enough to wash away cars and cover the land.

When lightning strikes the ground, it can cause fires and other damage.

Tornadoes Produce Strong Wind

Tornado

A tornado is a rotating column of air that extends from a cloud and moves along the ground. The greatest hazard from a tornado is destructive winds. The most violent tornadoes have winds of up to 300 miles per hour. The winds from a tornado are strong enough to pull trees out of the ground. Tornadoes can tear down homes and buildings. They can even pick up large objects and throw them from one place to another.

A tornado can destroy everything in its path.

The thunderstorms that form tornadoes can also produce large hail. Hail is balls of ice that fall from the sky. A ball of hail can be as small as a pea. It can also be as large as a softball. Large hailstorms can damage structures, such as the window of a car or the roof of a house, as well as crops.

Hurricanes Cause Wind and Flooding

People who live near a coast sometimes experience hurricanes. A hurricane is a large storm that forms over the ocean. It gathers more strength as it moves toward the land. Once it reaches the coast, its hazards include very heavy rain, large waves, rising water levels, and high winds.

The heavy rains and high water levels from a hurricane usually cause severe flooding that can wipe out roads and bridges. High winds from hurricanes damage houses, buildings, and other structures. An area that gets hit with a hurricane may be without power and water for several days or even weeks. It can take a very long time for a community to rebuild itself after a major hurricane.

The heavy rain and high winds of a hurricane can last for several hours.

Snow and Ice Are Winter Weather Hazards

Thunderstorms, tornadoes, and hurricanes mostly happen in warmer months. But extreme weather can cause hazards during colder months too.

One kind of extreme winter weather event is a blizzard. A blizzard is a winter storm that is very windy with a lot of snow. Blizzards can cause the roads to close. They can cause power outages. They can make it very difficult or even impossible for people to leave their homes to go to work or school.

An ice storm is another kind of winter weather. It occurs when precipitation freezes on roads, trees, and other surfaces. The ice is heavy. It can cause trees to snap. Power outages occur when ice builds up on power lines.

Some blizzards produce so much snow that they can bury cars!

Drought Causes Water and Food Shortages

All the extreme weather you have read about so far happens suddenly and is over with quickly. A **drought** is a long period of very dry weather. In areas that are usually dry, such as the desert, little precipitation is normal. But in areas that usually get a lot of rain, droughts can be hazardous. Droughts develop slowly and can last for months.

> **Vocabulary**
>
> **drought, n.** a long period of weather with less precipitation than normal

One hazard that is caused by droughts is crop damage. Crops cannot grow well without the water they get from rain. Farmers also have a hard time watering crops during a drought because the farmers must conserve, or save, water. This can lead to food shortages. Droughts can also lead to wildfires because the grass and trees in the area are so dry.

Drought can cause damage to crops that need water from rainfall.

Engineering for Extreme Weather

Chapter 7

Extreme weather hazards cannot be stopped. However, we can improve how we prepare for extreme weather hazards.

Engineers identify problems that occur because of hazardous weather. They use this information to design solutions, or ways to solve the problems. Solutions do not always work the first time. Sometimes they must be changed and improved after they are tested. This is all part of the **engineering design process**.

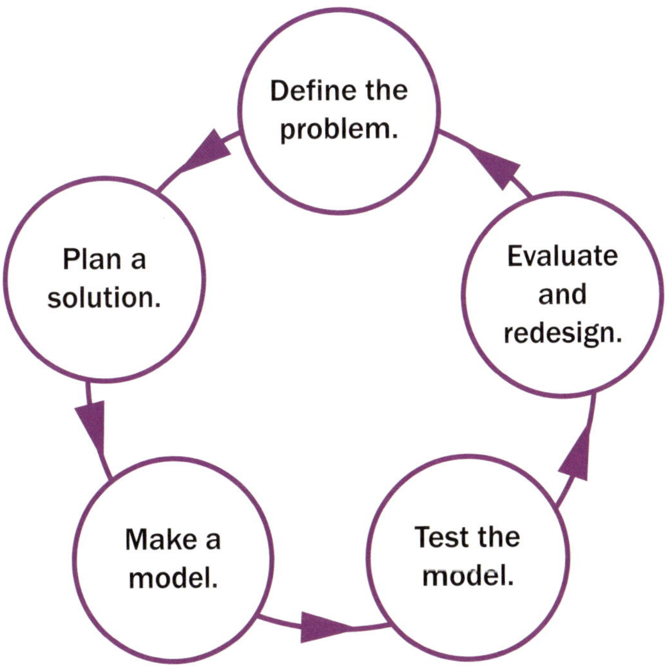

Big Question

How do engineers design solutions for extreme weather hazards?

Vocabulary

engineering design process, n. the steps that engineers take to solve a problem

Words to Know

In engineering, a *problem* is a situation that needs to be fixed in some way. A *solution* is a way of solving the problem.

Engineers Design Solutions for Flooding

Flooding can happen in any area that gets a lot of rain. But some places are more likely to flood than others. Engineers have designed many solutions to help prevent damage from floods.

Sandbags: Sandbags are bags of sand that are stacked around a structure to keep floodwaters out. People have been using sandbags for hundreds of years. They are inexpensive to make. But they have some disadvantages too. They only work in smaller floods. It takes a lot of time and work to make and stack sandbags. This means sandbags are effective only when a flood is predicted many days in advance.

Sandbags must be stacked a certain way to work correctly.

Levees and Seawalls: A levee is a wall that blocks water or forces it to move in a certain direction. Levees are made from both natural and artificial materials. Levees can break if there is too much water. Engineers find solutions to keep this from happening. They study how water flows. They apply what they learn to their designs.

During a hurricane, the sea rises quickly and can flood nearby land. Seawalls are structures that block this rising water. Engineers who design seawalls study an area's climate, coastline shape, and waves. These are the **constraints**, or limitations, of their designs. Then they design a seawall that will meet all the **criteria**, or requirements, that will make it successful. A successful seawall blocks water and prevents flooding.

Vocabulary

constraint, n. the limitation of a design

criteria, n. the requirements of a design for it to be a success

Without the protective seawall, these homes would flood more easily during storms.

Engineers Design Hurricane-Proof Buildings

Hurricanes bring destructive winds. These winds can destroy homes and buildings. People who live in hurricane-prone areas want to build structures that can withstand the winds. Some places have rules about how new buildings must be designed and built.

Engineers identify why some buildings cannot withstand hurricane winds. Then they find solutions to make buildings stronger and safer. They decide which materials work best. They investigate how the size and shape of a building can make it stronger or weaker. Then they build models to test their ideas. If a design does not meet all the criteria, or stand up against the wind in their tests, they come up with a new plan.

This hurricane-resistant home has a metal roof and sits up high on posts to avoid floodwaters.

Engineers Design Other Hazardous Weather Solutions

Lightning Rods: Have you ever seen a lightning strike that looked like it touched the ground? Lightning can strike objects on the ground. It can strike buildings too. Lightning strikes can cause fires and electrical outages. Tall buildings are struck more often because they are closer to the sky, but smaller buildings and homes can also be struck.

A lightning rod is a device that can be placed on top of a building to protect it from a lightning strike. A lightning rod is a skinny metal pole. It is designed to absorb the energy from the lightning. The rod moves the energy safely to the ground so that the lightning does not damage the structure that was hit.

Lightning rods redirect electricity to keep buildings and other structures safer.

Snow Fences: During a blizzard, wind causes snow to collect in large drifts. Wind often blows the snow onto roads. This can make roads unsafe. One solution to this problem is a snow fence. A snow fence is built along the sides of a road. Snow collects against the fence instead of drifting into the road.

Snow fences have been used for thousands of years. But their designs have improved over time. Engineers who design snow fences have to learn a lot about snow. They study how it blows and drifts in affected areas. They use this information to build models based on the data they collect. Then they evaluate the models to determine their effectiveness. Snow fences help communities save money on road salt and plows.

Word to Know

To *evaluate* something means to examine it and decide about its value. If you evaluate an answer to a question, you determine whether the answer is correct. If you evaluate a solution to a problem, you determine how well the solution works.

Automobile accidents on the road have decreased in areas that have snow fences.

Weather-Related Technology

Chapter 8

What do you do when you want to know what the weather will be like? You might watch a weather forecast. You might check a newspaper, website, or weather app. Forecasts are helpful. Often they are very accurate.

Big Question

Who are some inventors of weather-related technology?

Weather is not something that humans can control. It is impossible to know exactly what the weather will be like in the future. But people have invented weather-related technology to make forecasts more accurate. Technology is the use of scientific knowledge to make something helpful to others. People throughout history have designed technology to make it easier to observe, measure, predict, and deal with constantly changing weather. These inventors often follow the engineering design process.

Weather stations such as this one have different kinds of weather technology to help meteorologists observe and measure weather.

Word to Know

Technology is the use of scientific knowledge to make something that is helpful to others. Technology can be a device. It can also be a process, a good way to do something.

Evangelista Torricelli Invented the Barometer

Evangelista Torricelli was an Italian scientist, engineer, and mathematician. Torricelli invented a weather tool called a barometer. A barometer measures air pressure. Torricelli's barometer was a glass tube with mercury inside. The mercury level went up when the air pressure was high. It went down when the air pressure was low.

Measuring air pressure can help meteorologists predict weather. Changes in air pressure usually mean that the weather is changing too. The barometers that scientists use today look different than Torricelli's, but they measure air pressure to aid in accurate forecasting.

In the earliest barometers, liquid in a tube went up or down with changing air pressure.

Evangelista Torricelli lived during the 1600s.

Benjamin Franklin Invented the Lightning Rod

Do you remember reading about the lightning rod in the last chapter? A lightning rod is a skinny metal pole that is put on top of a building and that has an extension that goes into the ground. Lightning strikes the pole instead of the building. Then the electricity is carried safely to the ground.

Benjamin Franklin was an American scientist—among many other things! He was fascinated by thunderstorms. He learned through his experiments that lightning was a form of electricity. Franklin wanted to find a way to protect people and buildings from lightning strikes. He discovered that an iron needle would deflect lightning away from an object. He used this information to develop the lightning rod.

Benjamin Franklin studied lightning in the 1750s.

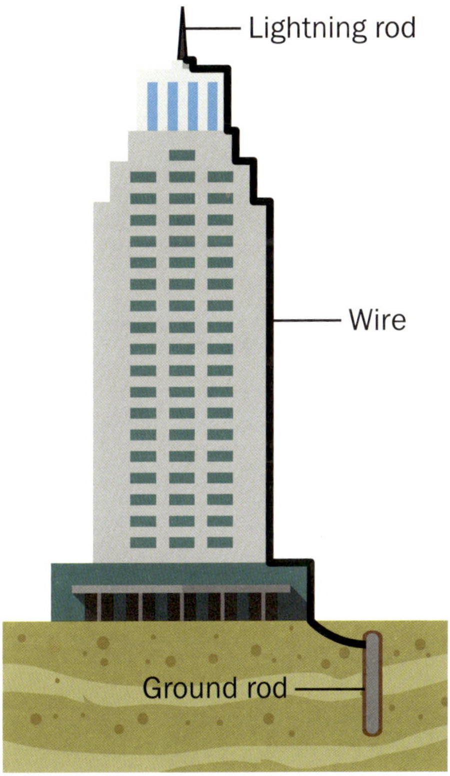

Lightning rods carry electricity from lightning to the ground through wires. This helps save buildings from damage.

Scientists Developed the Fujita-Pearson Scale

Not all weather technology is an object or tool that measures weather. The Fujita-Pearson scale is one example. This scale was invented in 1973 by two scientists, Dr. Tetsuya Fujita and Dr. Allen Pearson. It helps scientists classify the strength of a tornado based on the amount of damage it caused. It also considers the width of a tornado and the length of its path. The scale has six points. They range from F0 to F5.

Scientists have evaluated and improved the scale over the years. It is now referred to as the Enhanced Fujita Scale. It helps scientists collect and share valuable data about tornadoes.

Enhanced Fujita Scale			
EF 0	65–85 mph winds		
EF 1	86–110 mph		
EF 2	111–135 mph		
EF 3	136–165 mph		
EF 4	166–200 mph		
EF 5	>200 mph		

This is the Enhanced Fujita Scale that scientists use today. Wind speeds are calculated in miles per hour (mph).

Mary Anderson Invented Windshield Wipers

Windshield wipers do not help scientists measure, observe, or predict weather. But they do keep people safe in certain kinds of weather, such as rain and snow. Mary Anderson invented the windshield wiper in 1902. She was riding in a streetcar in New York City during a snowstorm. She noticed that the driver had to stop often to clean off the windshield by hand. She identified a problem and found a solution.

Mary Anderson thought of the idea for windshield wipers when there were still very few cars on the road.

Unfortunately, car manufacturers were not interested in this technology at the time. But Anderson's design led to the more modern versions of windshield wipers that we see today. In 2011 she was inducted into the National Inventors Hall of Fame.

Imagine not being able to travel by car whenever it rained because drivers could be unable to see clearly enough to move safely. Windshield wipers are now required on cars.

Joanne Simpson Studied Clouds

Joanne Simpson was the first woman to receive a PhD in meteorology, the study of weather. Simpson became interested in weather when she observed the clouds in the sky while she was sailing on a boat. As an atmospheric scientist, she continued to study clouds throughout her career.

Simpson was interested in the behavior of clouds during thunderstorms and hurricanes. She was the first person to create a cloud model, originally as a drawing. Later, she used computers to develop better models. Her cloud models helped scientists understand how clouds interact. This information could be used to give advance warning about whether a thunderstorm or other type of extreme weather was going to happen.

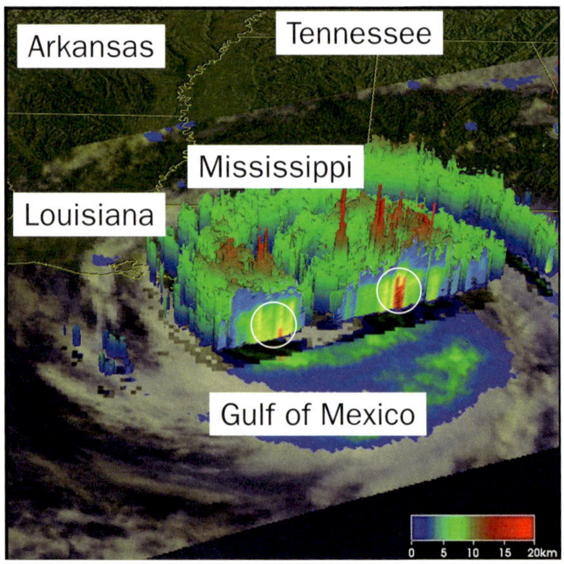

The TRMM is a satellite-based radar that Joanne Simpson helped develop. The satellite collects data. A computer turns the data into three-dimensional pictures of clouds. This image shows the movement of Hurricane Isaac over five hours. The red areas in the white circles are the eye of the storm at the start and at the end of the recording.

Simpson later helped develop the first radar that could measure rainfall in the tropics from space. This helped many scientists learn more about remote places on the planet. For example, it helped biologists better understand the rain forests.

Glossary

A

air mass, n. a large body of air in the atmosphere **(10)**

air pressure, n. the weight of air as it presses on objects below or within it **(3)**

atmosphere, n. the layer of air that surrounds Earth **(1)**

C

climate, n. the weather patterns in a place over a long period of time **(22)**

condense, v. to change from gas to liquid **(8)**

constraint, n. the limitation of a design **(33)**

criteria, n. the requirements of a design for it to be a success **(33)**

D

data, n. information that is observed or measured and recorded **(13)**

drought, n. a long period of weather with less precipitation than normal **(30)**

E

engineering design process, n. the steps that engineers take to solve a problem **(31)**

evaporate, v. to change from liquid to gas **(7)**

F

front, n. the place where two air masses meet **(18)**

H

hazard, n. a dangerous condition that can cause damage **(25)**

humidity, n. a measure of the amount of water vapor in the air **(7)**

M

meteorologist, n. a scientist who studies weather conditions and patterns **(13)**

P

precipitation, n. water that falls from the sky in the form of rain, snow, sleet, or hail **(6)**

prevailing winds, n. regular patterns of winds that blow from one direction **(12)**

W

water vapor, n. the gas form of water **(6)**

weather, n. what the air outside is like at any given time and place **(1)**

wind, n. the movement of air **(9)**

wind direction, n. the direction from which air moves when wind blows **(12)**

wind speed, n. a measure of how fast wind blows **(11)**

Core Knowledge®

CKSci™

Core Knowledge SCIENCE™

Editorial Directors
Daniel H. Franck and Richard B. Talbot

Subject Matter Expert

Scott Curtis, PhD
Distinguished Professor of Natural Sciences and Mathematics
Assistant Director, Center for Natural Hazards Research
Department of Geography
East Carolina University
Greenville, NC

Illustrations and Photo Credits

Andrew Cribb / Alamy Stock Photo: 35
Charles McClean / Alamy Stock Photo: 6
Emil Pozar / age fotostock / SuperStock: Cover A, 14b
F1 ONLINE / SuperStock: 9b
Fine Arts / Alamy Stock Photo: Cover C, 15c
Gary Whitton / Alamy Stock Photo: 23
GL Archive / Alamy Stock Photo: 41a
Grant Heilman Photography / Alamy Stock Photo: 15b
Jason Persoff Stormdoctor / Cultura Limited / SuperStock: 27a
Jeff Greenberg / age fotostock / SuperStock: 13
Jess Merrill / Alamy Stock Photo: 34
John Sirlin / Alamy Stock Photo: 8b
Jon Arnold Images / SuperStock: 4
Lars-Ove Jonsson / Alamy Stock Photo: 37
Loken / Mauritius / SuperStock: 41b
mario bonotto / Marka / SuperStock: 12b
MattX / Stockimo / Alamy Stock Photo: 8a
Mauritius / Mauritius / SuperStock: 32
MichaelGrant / Alamy Stock Photo: 36
NASA Image Collection / Alamy Stock Photo: 42
olrat / Alamy Stock Photo: 30
Pantheon / SuperStock: 24b
Panther Media GmbH / Alamy Stock Photo: 2
Peter Scholey / Alamy Stock Photo: Cover B, 29
Photononstop / SuperStock: 10a
Rainer Lesniewski / Alamy Stock Vector: i, iii, 17b, 18
RGB Ventures / SuperStock / Alamy Stock Photo: 5
RooM the Agency / Alamy Stock Photo: Cover D
Ryan McGinnis / age fotostock / SuperStock: 27b
Science and Society / SuperStock: 38b
Science Photo Library / SuperStock: 14a, 15a
Stocktrek Images, Inc. / Alamy Stock Photo: 11
studiomode / Alamy Stock Photo: 14c
Super Nova Images / Alamy Stock Photo: 28
Susan E. Degginger / Alamy Stock Photo: 22
Tetra Images / SuperStock: 9a
Universal Images / SuperStock: 38a
Victor de Schwanberg / Alamy Stock Photo: 33
Westend61 / SuperStock: 12a
Wim Wiskerke / Alamy Stock Photo: 39a
Yury Dmitrienko / Alamy Stock Photo: 26
Zoonar GmbH / Alamy Stock Photo: 1